Trailside Plants of Hawai'i's National Parks

By Charles H. Lamoureux
Director, Harold L. Lyon Arboretum and
Professor of Botany, University of Hawai'i at Mānoa

HAWAI'I NATURAL HISTORY ASSOCIATION

Published by the
Hawai'i Natural History Association
in cooperation with the
National Park Service
U. S. Department of the Interior

ISBN 0-940295-09-1

2

Trailside Plants of Hawai'i's National Parks

By Charles H. Lamoureux
Director, Harold L. Lyon Arboretum and
Professor of Botany, University of Hawai'i at Mānoa

Preface

Hawai'i has many plants which are found only here and nowhere else in the world, and many others which are restricted to the tropics. Visitors from temperate areas will probably find most Hawaiian plants to be new to them.

About 1,000 different kinds of higher plants (flowering plants, ferns and fern allies) have been recorded as growing within the parks' boundaries. Many of these are uncommon, some are very rare, others grow only in remote areas, and others, although fairly common, are inconspicuous and unlikely to be noticed by the casual visitor. This book covers some of the more common and conspicuous plants, the ones any visitor who walks a few of the trails of Hawai'i Volcanoes National Park or Haleakalā National Park is likely to encounter. Unless otherwise indicated, the locations given for plants are within Hawai'i Volcanoes National Park.

For each plant a brief description is provided which, along with the photograph, should enable one to identify the plant. Places where the plant can be found readily are indicated. Also, brief mention is made of the ways in which the plants were used by the Hawaiian people, if such uses have been recorded.

The scientific and Hawaiian names of flowering plants used here are those used by Wagner, Herbst and Sohmer in the *Manual of the Flowering Plants of Hawai'i*.

For the reader who would like more information about the plants in this book, as well as many additional species, Stone and Pratt's *Hawai'i's Plants and Animals: Biological Sketches of Hawaii Volcanoes National Park* is recommended.

3

Hawaiian Plants

Origin and Evolution

4 The Hawaiian Islands were all formed by volcanoes which arose from beneath the sea as barren sterile rocks. The islands have always been isolated from continental land masses. Consequently, the only plants which grew here before the coming of humans were those descended from ancestors that had been able to cross at least 2,000 miles of ocean. These ancestral plants were ones which had some mechanism that permitted long-distance dispersal. Some had spores or seeds which were light enough to be carried by wind. Others had seeds which were able to float on sea water, and which had embryos within the seeds that were not killed by contact with salt water. Still other types were carried here by birds. None of these methods is very efficient, but botanists have estimated that about 275 successful introductions of flowering plants and 135 successful introductions of ferns and fern allies could account for all the native higher plants known from Hawai'i. Geological evidence suggests that the oldest islands in the Hawaiian chain were above sea level at least 27 million years ago and perhaps as much as 70 million years ago. Thus an average of one successful new immigrant flowering plant every 100,000 years and one successful new immigrant fern every 200,000 years would be sufficient to give rise to the known native Hawaiian flora.

When these new immigrants arrived and became established and spread into new areas, they changed gradually with time — they evolved. Evolution proceeds by natural selection. Those plants best adapted to a particular environment are the ones which have the best chance of surviving and reproducing in that environment.

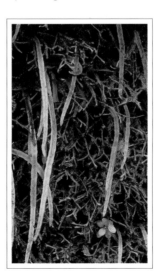

Just as very few plants were able successfully to cross 2,000 miles of ocean in reaching Hawai'i, very few animals were able to reach here without human help. Birds and several kinds of insects made the trip successfully, as did a few snails, but the only mammal was a bat. There were no cattle, sheep, goats, deer, pigs, rats, cats, or dogs. While plants were evolving in Hawai'i there were no large mammals grazing or browsing on them, and thus, the plants did not evolve defense mechanisms which would enable them to withstand such pressures.

In time many new kinds of plants and animals evolved here. Biologists use the term *native* to refer to plants or animals which live in an area without having been brought there by humans, and use the terms *alien*, *introduced* or *exotic* to refer to those brought into an area by humans, either intentionally or unintentionally. Among native species there are further distinctions made between *endemic* species (those native to a small geographic area such as Hawai'i), and *indigenous* ones (such as a species which is native to New Guinea, Fiji, Samoa, and Hawai'i).

Most of the native plants and animals in Hawai'i have evolved to the point that they are recognized as different species from their ancestors and from species elsewhere in the world. According to the most recent summary given by Wagner, Herbst and Sohmer, the native Hawaiian flora consisted of 1,094 species, subspecies, and varieties of flowering plants, of which 91% were endemic. In other words, 91% of all Hawaiian flowering plants grew only in Hawai'i and no place else on earth.

When the first Polynesian settlers arrived in Hawai'i, nearly 2,000 years ago, they brought with them about 25 species of plants which were used for food, fiber, or medicinal purposes. They also brought the pig, dog, jungle fowl, and rat. All of these must have had some effect on the original flora and fauna, but we are only just starting to learn how great these effects were.

Since 1778, the islands have been exposed to sustained contact with the rest of the world, and great changes have occurred in the Hawaiian environment, which have led to the disappearance of many of Hawai'i's unique plants and animals. Early sea captains introduced goats, sheep, and cattle, which have had especially severe effects on the "defenseless" Hawaiian plants that had evolved in the absence of such animals. At the same time, new plants were being introduced, and many of these spread into areas where the native plants were being destroyed by grazing. Extensive areas of land have also been cleared for agriculture and housing.

When the native plants disappeared, the native birds, insects, and land snails which lived and fed on them

disappeared also. There are no complete tabulations, but we know that many plant species have become extinct in the past 200 years, and we estimate that from one-half to one-third of the native plants are now rare and probably endangered.

It is only in places like National Parks and nature reserves that unique Hawaiian plants and animals are likely to be conserved. Even within Haleakalā and Hawai'i Volcanoes National Parks there has been widespread destruction in the past by goats, pigs, and cattle. Much work is now going on to control remaining goat and pig populations and alien plant pests, and to re-establish rare plants in areas where they were formerly more abundant. Without these vigorous efforts by the National Park Service, even some of the common native plants illustrated in this booklet could disappear in another generation.

Club Moss Family

Lycopodiaceae
Lycopodium cernuum; wāwaeʻiole

Wāwaeʻiole is often found growing in thickets with uluhe. Look for it between Hawaiʻi Volcanoes' Park Headquarters and Steaming Bluff. The plant consists of long branching stems closely covered with small scale-like leaves. The main stem trails along the ground and forms root clusters at intervals, but side branches are erect. Small cones, which form at turned-down branch tips, give rise to spores by which the plants reproduce. These also were the inspiration for the Hawaiian name, which means "rat's foot."

Today, wāwaeʻiole is harvested from Hawaiian forests for use in flower arrangements. Once, Hawaiians reportedly used the plant as a treatment for rheumatism, by having the patient bathe in water in which wāwaeʻiole plants had been boiled for several hours.

Gleichenia Fern Family

Gleicheniaceae
Dicranopteris emarginata,
D. linearis;
uluhe, false staghorn fern

The uluhe ferns form nearly impenetrable thickets in open areas throughout wetter parts of Hawai'i Volcanoes. They are abundant in the area between Volcano House, Sulphur Bank, and Steaming Bluff.

The plants have long creeping underground stems which produce fronds ten to fifteen feet long. These fork repeatedly and produce the dense tangles of vegetation which become three or four feet deep, and even climb up into the lower branches of trees. These tangles develop quickly when openings are created in forests, but the plants do not grow well in deep shade.

Two species of uluhe occur in Hawai'i Volcanoes.
D. emarginata, above, which has the undersides of the leaves covered with small rusty hairs, is the more common species, and is endemic to the Hawaiian Islands. *D. linearis,* which has smooth leaves lacking rusty hairs, is native to Hawai'i and also to many other Pacific islands and tropical and sub-tropical Asia.

Tree Fern Family

Dicksoniaceae
Cibotium glaucum; hāpu'u
C. chamissoi; hāpu'u 'i'i
C. hawaiense; meu

Tree ferns are a very conspicuous part of the vegetation of Hawai'i Volcanoes, for they form a nearly continuous understory in wetter forests. The fern jungle on Crater Rim Drive is a popular photographic subject. The larger tree ferns are these three species of *Cibotium*, but some species of *Sadleria* also become small tree ferns.

Cibotium tree ferns develop large trunks from six to eight to over twenty feet tall and from one to three feet or more in diameter. Most of the trunk consists of dark-colored interwoven roots which surround a small central stem three or four inches in diameter. These tightly-woven root masses make a good potting medium for growing orchids, and many tree ferns are harvested outside Hawai'i Volcanoes for this purpose.

At the top of the trunk is a cluster of fronds each six to twelve feet long and divided into hundreds of small segments. Young fronds, the bases of older fronds, and the bud at the top of the trunk, are densely covered with silky scales or bristles called pulu. For the Hawaiians, pulu was used as an absorbent surgical dressing, and in embalming the dead. From about 1860 to 1885, pulu was harvested commercially for use as pillow and mattress stuffing. Over four million pounds were shipped during this period, mostly to California. The pulu factory was located near Nāpau Crater on land which is now within the park.

The pulu of hāpu'u, opposite page, is soft, silky, and yellowish brown, whereas the pulu from hāpu'u 'i'i, lower left, consists of coarse, stiff, reddish brown bristles. Both species are much more common than the meu, which is uncommon in Hawai'i Volcanoes except in the 'Ōla'a tract. The trunk of the meu, below, is taller and thinner than the trunk of the other species, and its fronds, immediate left, are smaller.

Davallia Fern Family

Davalliaceae
Nephrolepis exaltata; niʻaniʻau,
ʻōkupukupu, common swordfern
N. cordifolia; pāmoho, narrow
swordfern
N. multiflora; scaly swordfern

These three ferns are found throughout the lower and middle elevations of both parks. Common swordfern, upper left, and scaly swordfern, above, are among the first plants to colonize new lava flows, and are common in cracks in lava, near steam vents, and in other open places. Common swordfern is more widely distributed, but scaly swordfern is more common at lower elevations, such as in Hawaiʻi Volcanoes' Kalapana section. Narrow swordfern, below left, is usually found in wet forests and is distinguished by fronds less than two inches wide. The fronds in the other species are usually over two-and-one-half inches wide. The lower leaf surface of the scaly swordfern, an introduced species, bears many small, brown scales giving the surface a "cobwebby" appearance, whereas the native common swordfern lacks these.

Blechnum Fern Family

Blechnaceae
Sadleria cyatheoides; 'ama'u
S. pallida; 'ama'u

Sadleria is a genus of ferns endemic to the Hawaiian Islands. Two species are common in both parks. Both species develop short trunks, usually one to two feet and rarely up to nine feet high and become small tree ferns, although they rarely reach the size of hāpu'u tree ferns. The fronds of both species are frequently red when new and green at maturity.

S. cyatheoides, page 12, page 13 top and page 13 lower left, is generally the larger of the two species and is more widely distributed, ranging from open lava flows to wet forests. It is very common in and around Hawai'i Volcanoes' Kīlauea Crater, including near Halema'uma'u,

the firepit of Kīlauea, "the home of the 'ama'u." In Haleakalā it grows along the Halemau'u Trail on Leleiwi Pali, and on the palis behind Palikū.

S. pallida, page 13 far right, is usually a smaller plant and is

confined to wetter places, such as the forest near Hawai'i Volcanoes' Thurston Lava Tube, and in Haleakalā's Kīpahulu Valley.

The species can be readily separated by examining the lower sides of the frond segments. In *S. cyatheoides* each segment is uniformly light green with a single prominent midvein. In *S. pallida* each segment has many prominent small side veins, which appear translucent when held up to the light.

The Hawaiians obtained a red dye from the outer part of the trunk, and leaves were occasionally used for thatching. Young leaves and the pith of the stem were sometimes used as food, but apparently only when more desirable food was unavailable. In recent years a weevil which bores into the frond stalks of these plants has become a serious pest in many parts of Hawai'i Volcanoes, and many plants have damaged fronds.

12

Screwpine Family

Pandanaceae
Pandanus tectorius; hala

The hala is a tree found in moist coastal areas, as at Hawai'i Volcanoes' Kalapana section and Haleakalā's 'Ohe'o Gulch. The trees have stiff, stilt-like prop roots which extend at steep angles from the stem. Stems and branches bear conspicuous rings which are the scars of fallen leaves. The leaves are generally from three to six feet long and two to four inches wide, with prickly margins, and grow in spiral clusters at branch tips. The trees are of separate sexes. Male trees produce large, fragrant, white clusters of pollen-bearing flowers called hīnano whereas female trees bear the spherical clusters, composed of many wedge-shaped, woody fruits.

These turn yellow, orange or red on ripening as they fall from the clusters. These fruits, on drying, can float for long periods, and have probably been the means by which *Pandanus* has spread from island to island in the Pacific.

The leaves (lau hala) were used by the Hawaiians as their major weaving material for mats, baskets, and other household objects. They were also employed as thatch. Today they are used for many craft items. In other parts of the Pacific the ripe fruits were eaten, and plants were selected and cultivated for their desirable edible properties. The Hawaiians seem to have preferred different foods and did not select hala plants for their edibility. Consequently, hala fruits in Hawai'i are not very palatable. The fleshy bases of fruit segments are occasionally strung into colorful leis.

Grass Family

Poaceae
Andropogon virginicus;
broomsedge
Schizachyrium condensatum;
bushy beardgrass

These two grasses, abundant in Hawai'i Volcanoes today, are species introduced to Hawai'i by humans in recent years. Broomsedge, native to the eastern United States, was first recorded in Hawai'i in 1932, and bushy beardgrass was introduced from tropical America around 1960.

Both species are common below the 4,000 foot elevation, especially along Crater Rim Drive and Chain of Craters Road. They are unwelcome invaders, for they grow in dense masses which inhibit the growth of seedlings of native species. Perhaps even more importantly, their dry dead stems and leaves do not decompose readily but remain erect and create a considerable fire hazard — one of the dangers to rare native plants. Both species grow in areas which formerly supported only scattered trees and shrubs.

These are similar looking bunch grasses which reach heights of three to four feet, but differ in the nature of the flowering spikes. In broomsedge, foreground, these are narrow and cylindrical whereas in bushy beardgrass, background, they are shaped like inverted cones, narrow at the base and broadest at the top.

Grass Family

Poaceae
Heteropogon contortus; pili

This grass is common in drier sections of Hawai'i Volcanoes, and it forms the dominant grass cover in parts of the Kalapana section.

It is a bunch grass, up to three feet tall, which can be recognized by its bluish green leaves, and its narrow flower spikes. Each spikelet is tipped with a twisted reddish brown awn or bristle about four inches long.

Pili was the favored thatching material of the Hawaiians, and most houses were thatched with this plant when it was available.

Sedge Family

Cyperaceae
Machaerina angustifolia; 'uki

18

The 'uki is a coarse sedge with sword-shaped leaves two to three feet long and up to one inch wide. Plants grow in clumps which are common in open forests around Kīlauea Crater and along the Chain of Craters Road.

Flower stalks three to four feet tall bear clusters of tiny flowers, and later fruits, which are surrounded by small brownish to blackish bracts. The bracts remain attached to the stalk indefinitely, which makes the flower stalks suitable for use in dry floral arrangements. Hawaiians used 'uki leaves to line the inner walls of their grass houses.

Lily Family

Liliaceae
Astelia menziesiana; pa'iniu

This species of pa'iniu is endemic to the Hawaiian Islands. In Hawai'i Volcanoes it grows best in the wet forests near Thurston Lava Tube, but it is sometimes found in fairly open areas such as along Devastation Trail. It also grows in Haleakalā, mainly in the wet forests.

The plants may grow as epiphytes on the branches and trunks of trees, or they may be rooted in the ground. The leaves, which grow in a rosette, are up to two feet long and one to two inches wide. These are a bright silver since they are covered with shiny grayish white scales. Clusters of small greenish flowers, male and female flowers on separate plants, develop on stalks which grow out from the center of the rosette. The female flowers form bright orange berries, which are eaten by birds. In this way the seeds are spread.

Agave Family

Agavaceae
Cordyline fruticosa; tī

The tī was brought to Hawai'i by early Polynesian immigrants and was widely grown by the Hawaiians. Today it is common in cultivation, and persists in the wild in many areas formerly cultivated. Look for it especially in Hawai'i Volcanoes' Kalapana section and Kīpuka Puaulu, and at Haleakalā's 'Ohe'o Gulch. Plants have unbranched or sparingly-branched woody stems, up to 12 feet tall, bearing large clusters of leaves at their tips. Each leaf is 12 to 24 inches long and up to 4 inches wide, with a shiny green surface. (Tī plants with purple, red, pink, or striped leaves belong to horticultural varieties which were brought to Hawai'i only after European contact). Large clusters of many tiny white or purplish flowers form at the branch tips, but in wild green tī they never form fruits.

Tī was a most useful plant to the Hawaiians. Leaves were used to wrap food, and as plates. They were employed as thatch for houses, woven into sandals for crossing 'a'ā lava flows, tied to the long hukilau nets, and later, fashioned into skirts worn by hula dancers. A branch tip with its leaf cluster makes an impromptu sled for use on a muddy forest slope, and children of all ages still gather at tī leaf slides during heavy rains. The roots store large amounts of sugar, and were baked and eaten both as a confection and as an emergency food in times of famine. The brandy-like 'ōkolehao is distilled from tī roots which have been baked, mashed, and fermented.

The tī plant was a symbol of divine power to the people of old, and was considered a charm against unwitting transgression of a tabu. Hawaiian priests wore tī leaves around their necks, and leaves often formed part of religious offerings. Even today, many people place a tī leaf in the hole in which a tree is about to be planted.

Iris Family

Iridaceae
Crocosmia x crocosmiiflora;
montbretia

This plant, with its conspicuous clusters of brilliant orange flowers, is readily noticed where it grows along roadsides and in disturbed places in the Kīlauea area.

It is a hybrid, the result of a cross between two South African plants, *Crocosmia pottsii x C. aurea*, which has escaped from gardens and become naturalized in Hawai'i. Like many hybrids this plant is sterile; it does not produce seeds. However, it has short thick underground stems, or corms, which bud off many smaller corms and these spread the plant rapidly. It is an undesirable weed in the park, where it crowds out native plants.

Ginger Family

Zingiberaceae
Hedychium coronarium;
white ginger, 'awapuhi ke'oke'o
H. flavescens; yellow ginger,
'awapuhi melemele
H. gardnerianum; kāhili ginger

Three ornamental species of ginger have escaped from cultivation and become established in Hawai'i Volcanoes. Each has stems three to six feet tall which bear two rows of leaves up to eighteen inches long and three to four inches wide.

Yellow ginger, left, and white ginger, right, are similar, and both grow around the 1877 Volcano House near Park Headquarters.

Kāhili ginger, above right, has fragrant yellow flowers in foot-long inflorescences at stem tips. The inflorescence resembles a Hawaiian kāhili, hence the common name. Look for it near Volcano House and on the Halema'uma'u Trail. Birds spread its seeds, and it is rapidly becoming a major weed in rain forests at Kīlauea.

Orchid Family

Orchidaceae
Arundina graminifolia;
bamboo orchid

24 Although Hawai'i is famous as a world center of orchid culture, the orchids grown commerically here are all plants which have been brought in by humans from the tropical areas of Asia, Africa, and America. In fact, Hawai'i has only three species of native orchids, fewer than any other state. Furthermore, of the hundreds of kinds of orchids now grown in Hawai'i only four have become wild here.

The three native orchids are all plants of wet rain forests with small, inconspicuous flowers. As far as we know, they were never given Hawaiian names. Two of them grow within Hawai'i

Volcanoes in wet forests, but they are rare here. They are more common in the Kīpahulu section of Haleakalā.

Three of the four introduced species which have become naturalized occur in the park.

The Chinese ground orchid, *Phaius tankarvilleae*, is found occasionally in wet forests, such as those near Thurston Lava Tube. The Philippine ground orchid, *Spathoglottis plicata*, with small purple flowers on three-foot stalks, is widely scattered within the park. The bamboo orchid, *Arundina graminifolia*, left, is common in open grassy areas near Sulphur Banks and Steaming Bluff.

These are up to six feet tall, with narrow grass-like leaves. The purple and white flowers are short-lived and only one or two flowers are usually open at the same time.

Pepper Family

Piperaceae
Peperomia ssp.; 'ala'ala wai nui

At least ten different species of *Peperomia* grow in the two parks. All are small herbs with fleshy stems and thick succulent leaves. Tiny flowers and sticky fruits develop on club-like spikes at stem tips.

Among the more common species are:

Peperomia leptostachya — this species is found growing in dry to mesic sections of both parks, mostly below 2,000 feet elevation.

P. cookiana — grows in large patches along the trail in Kīpuka Puaulu in Hawai'i Volcanoes, and Kaupō Gap in Haleakalā. The leaves are often red beneath.

P. tetraphylla — are small plants, growing on rocks or trees, with thick succulent leaves borne in whorls of four. It is found in a wide variety of habitats.

Hawaiians used *Peperomia* to compound medicines for a host of ailments, including general debility and pulmonary diseases.

Nettle Family

Urticaceae
Pipturus albidus, P. forbesii;
māmaki

The māmaki is a large shrub or small tree found in open forests throughout both parks from sea level to the 5,000 foot elevation. *Pipturus albidus* is particularly common in Hawai'i Volcanoes' Kīpuka Puaulu and *P. forbesii* in Haleakalā in Kaupō Gap south of Kuiki.

Leaves are ovate, light green above and whitish beneath. Leafstalks and veins are sometimes red, sometimes green. The flowers occur in small clusters in leaf axils, with male and female flowers on separate plants. Female flowers develop into peculiar fleshy, waxy white, compound fruits up to one inch in diameter.

The fruits were occasionally used as medicine, but the major use of the plant was in the production of kapa or bark cloth.

Māmaki is the major host plant for the larvae of the Kamehameha butterfly, one of only two known native Hawaiian butterflies. The caterpillar's head mimics the white knobby fruit cluster of the māmaki.

Sandalwood Family

Santalaceae
Santalum paniculatum, S. haleakalae;
'iliahi, sandalwood

The sandalwoods in both parks form small trees, usually 15 to 20 feet tall and up to 6 inches in diameter. Branches bear opposite, light green, leathery leaves. In the Hawai'i Volcanoes species, *S. paniculatum*, center, clusters of small greenish flowers form at branch tips and develop into bluish black fruits about one-half inch long, each with a single large stone. It is fairly common just west of Kīlauea Military Camp and on the upper Hilina Pali Road. The Haleakalā species, below, is distinctive for its deep red flowers. It grows above the park entrance, in Ko'olau Gap, and along the Kaupō Trail below Palikū.

There are four currently recognized species of sandalwoods native to Hawai'i. During the first third of the 19th Century, vast quantities of the fragrant heartwood were shipped to China as the first major export item of the Hawaiian Islands. This trade was such an important part of life in Hawai'i that this period of Hawaiian history is sometimes called the "Sandalwood Period." Before this time there must have been extensive groves of sandalwood, given the amount of wood exported, but exploitation was so great and sandalwood trees became so reduced in numbers that today there are no large groves anywhere in the islands.

Most sandalwoods are reported to be root parasites. Their roots form underground connections with the roots of other plant species from which they obtain substances needed for growth.

Rose Family

Rosaceae
Fragaria vesca, F. chiloensis ssp.
sandwicensis;
'ōhelo papa, strawberry

Strawberries are plants which love cool climates, and so on the islands of Hawai'i they are found between 3,500 and 6,000 feet in moderately wet forests. When the red-fruited *Fragaria chiloensis* was common it was eaten by the nēnē and by Hawaiians. Now, this native is rare, but it can still be found in Haleakalā.

The introduced *F. vesca*, above, a native of Europe and North America, looks very similar, but its leaves are thinner, the undersides of leaves are less hairy, and the flowers and fruits rise above the leaves. Its fruits can be either red or white. Look for it in Haleakalā's Hosmer Grove, and Hawai'i Volcanoes' Kīpuka Puaulu and Kīpuka Kī, where the white-fruited form grows.

Rose Family

Rosaceae
Osteomeles anthyllidifolia; 'ūlei

The 'ūlei is indigenous to the Hawaiian Islands, and is also found in Tonga and the Cook Islands. It grows in the dryer parts of both parks. In Hawai'i Volcanoes it is found from sea level to about 4,000 feet elevation. Look for it along the road to Kīpuka Nēnē. It grows in Haleakalā Crater below 7,500 feet.

In open rocky areas it is a low creeping or spreading shrub, forming a dense ground cover a few inches high. In more protected areas, on better soils, it grows into a large erect shrub or even a small tree. The pinnate leaves, each composed of a number of oblong leaflets, are silvery because of a dense coating of silky hairs. White flowers, about one-half inch in diameter, are clustered at branch tips. Its fruits are white, fleshy berries.

The Hawaiians used the strong, flexible woody stems for fish spears and digging sticks, as hoops to hold open the mouths of fish nets, and for a musical instrument called 'ūkēkē.

Rose Family

Rosaceae
Rubus hawaiiensis;
'ākala, Hawaiian raspberry
R. argutus; prickly Florida
blackberry
R. rosifolius; thimbleberry
R. ellipticus; yellow Himalayan
raspberry

30 The 'ākala, this page and page 31 lower right, is an endemic species which looks much like a large raspberry, with canes up to 12 feet long and large hairy leaves. The stems are usually without prickles. Flowers are pink, up to one-and-one-half inches in diameter. The berries are salmon pink to red and very large, as much as two inches in diameter, and are quite juicy. Some are sweet, others have a slightly bitter taste. The Hawaiians used to eat the berries, and obtained a dye from them. Bark cloth (kapa) was occasionally made from the stems.

The prickly Florida blackberry was introduced to Hawai'i about 1904 from the southeastern United States, and has become a major pest in the islands. In Kīpuka Puaulu it forms large thickets which are almost impenetrable because the plants are covered with strong prickles. Flowers are white, fruits black.

In places around Kīlauea the Himalayan raspberry, *Rubus ellipticus,* has also become naturalized and is a pest. Its ripe fruits are yellow.

The thimbleberry, page 31 upper left, was introduced from tropical Asia in the 1880s and has spread widely throughout the islands, especially in wet forests. Plants are up to three feet tall and are covered with small prickles, but the plants usually grow separately and do not form the dense tangles which blackberries do. The flowers are white and fruits are red, the size and shape of an ordinary

raspberry, with a slightly sweet but somewhat insipid taste.

The 'ākala, thimbleberry, and prickly Florida blackberry are found in Kīpuka Puaulu of Hawai'i Volcanoes. The yellow Himalayan raspberry is common in the 'Ōla'a Tract. In Haleakalā Crater the 'ākala grows at Palikū and surrounding palis, as does blackberry. Thimbleberry is common in Kīpahulu Valley. Another endemic 'ākala, *Rubus macraei,* is less common. It may be found at Hosmer Grove and other places in subalpine shrubland on Haleakalā. It is a small creeping plant with dark red to purple fruits.

Pea Family

Fabaceae
Acacia koa; koa

The koa is the second most common tree in Hawai'i Volcanoes, and one of the largest native Hawaiian trees. Trees 100 feet tall or more and up to 6 feet in diameter grow in dense forests east of the park at about 5,000 feet elevation. Within the park it occurs from about 1,500 to 7,000 feet and is observed most easily in Kīpuka Puaulu and along the Mauna Loa Road. In Haleakalā, a different variety of koa is easily found in Kaupō Gap, and is abundant in Kīpahulu Valley.

The leaves on mature trees are not true leaves but sickle-shaped flattened leaf stalks called phyllodes. In the Haleakalā type these are half the width of the Big Island type. True leaves, page 32 top, which are pinnately divided, occur on seedlings, root sprouts, and occasionally toward the bases of new branches. The yellow flowers appear in small globular heads of bloom, replaced as they mature by flattened brown seed pods up to six inches long and one inch wide.

Koa reproduces both by seedlings and by root sprouts. In the mountain parkland on the Mauna Loa Road there are groves of koa, up to half an acre in extent, with a large old tree in the center and gradually smaller trees toward the edges. Each such grove is probably a single colony formed of root sprouts from the original tree.

The wood was highly prized by the Hawaiians, and was used for canoes, calabashes, and surfboards. It is still the most valuable timber tree in Hawai'i, valued for its beautiful finish.

Pea Family

Fabaceae
Sophora chrysophylla;
māmane

34 The māmane is a tree distributed widely within both parks from about the 1,500 foot elevation to the tree line at about 8,500 feet. Like many other Hawaiian plants, it has several forms. Mature plants may be erect shrubs a few feet high or trees 40 feet tall. The leaves are composed of separate leaflets and are usually yellowish green when young, becoming grayish green as they mature. The bright yellow sweetpea-like flowers, each about one inch long, develop in clusters at branch tips, and a tree in full flower is a spectacular sight. Heaviest flowering occurs during winter months.

Winged seed pods, up to six inches long, are first green, later becoming brown and woody.

They remain on the tree for several months before splitting and releasing the small yellow seeds. Many native birds feed among the flowers, and on Mauna Kea the young seed pods are a major item in the diet of the rare palila.

Scattered trees are present near Hawai'i Volcanoes' Park Headquarters and along the Chain of Craters Road, but māmane is especially common in Kīpuka Nēnē, Kīpuka Puaulu, and along the Mauna Loa Road. Māmane is widespread below 8,500 feet elevation within Haleakalā Crater, and on the rim above park headquarters. Goats, sheep and cattle feed eagerly on the foliage and on young plants, and māmane is unable to reproduce when large numbers of browsing animals are present. Without the protection offered by National Parks the tree could soon become rare.

The Hawaiians used the wood of māmane for digging sticks and hōlua sled runners. Today it is the favored wood for fence posts.

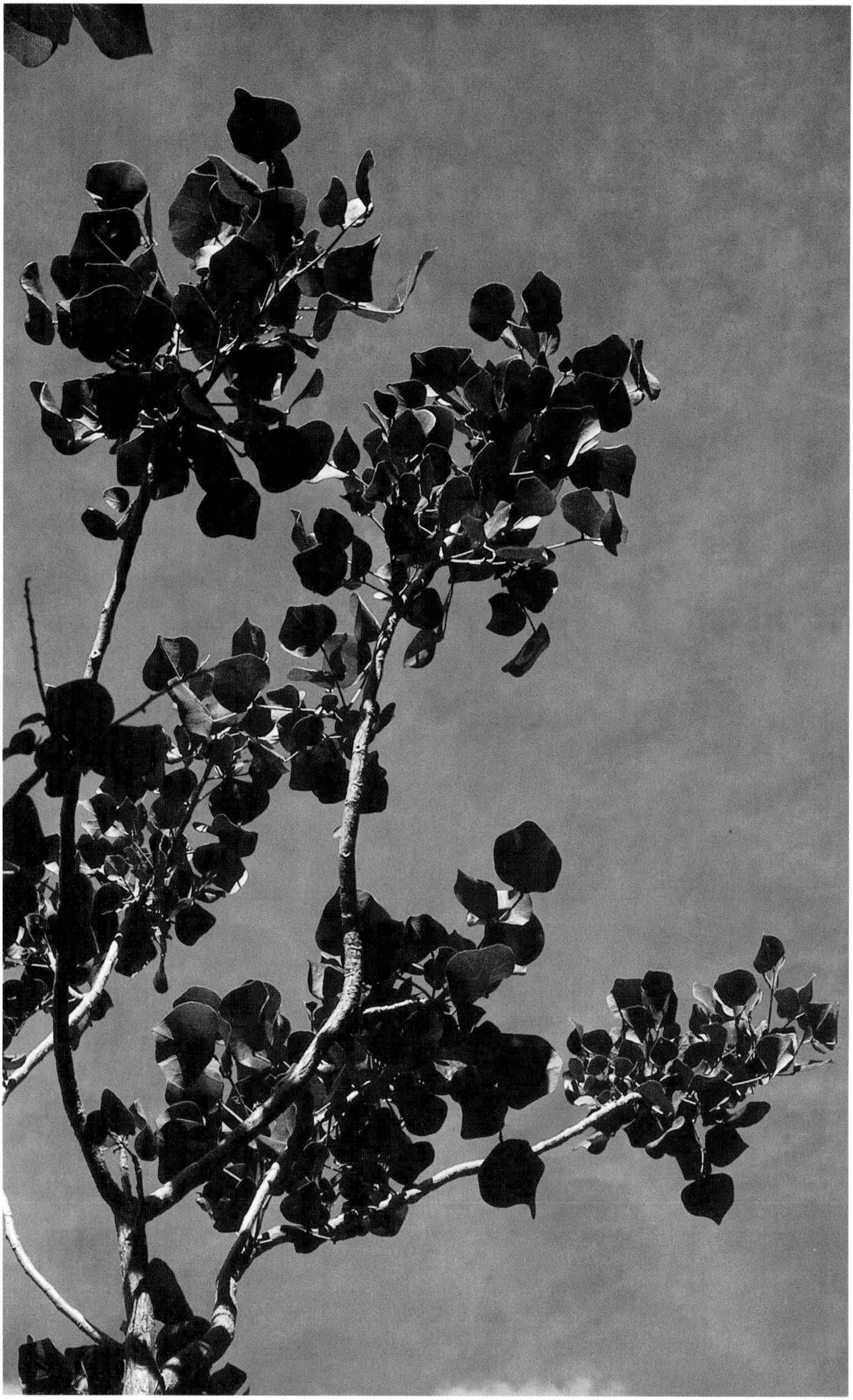

Pea Family

Fabaceae
Erythrina sandwicensis; wiliwili

The wiliwili is an endemic Hawaiian species, closely related to the coral trees used as ornamental trees throughout the tropics. They are confined to the dry lowlands such as in the Kalapana section of Hawai'i Volcanoes.

This is one of the few Hawaiian plants which is deciduous. Between July and September, during the driest part of the year, the trees lose all their leaves. While leafless, the clusters of colorful flowers open. Plants in the park have flowers which are mostly orange or greenish orange in color, but other colonies in Hawai'i may also have flowers that are red, yellow, green, or even white. The brown, hairy pods contain a few red bean-like seeds which are sometimes used to make leis.

Trees grow to thirty feet tall or more, with trunks often three feet in diameter. These are sometimes straight, but more often they are distinctively twisted and gnarled. The bark is a rusty brown color with greenish streaks. The wood is soft and very light in weight, and was used to make canoe outriggers and fish net floats.

Pea Family

Fabaceae
Senna surattensis; kolomona

Kolomona is only one of several introduced species of *Senna* within Hawai'i Volcanoes, but it is one of the most common. This shrub grows six to twelve feet high on poor lands in the coastal forest. Its clusters of yellow flowers with whitish green leaves beneath make it conspicuous. The genus *Senna* is a very large group, with more than 250 species spread across tropical and warm lands.

Geranium Family

Geraniaceae
Geranium cuneatum ssp. hypoleucum;
G. c. ssp. tridens, G. multiflorum,
G. arboreum; hinahina
G. homeanum; cranesbill

Scattered above 5,000 feet on Maui, Hawai'i and Kaua'i are six woody members of the geranium family. Like silversword, these have adapted to the high elevation heat by acquiring through evolution the fine white hairs which reflect sunlight. The underside of each leaf is silvery.

G. cuneatum ssp. hypoleucum, above, is the common type from 7,000 to 8,000 feet along Hawai'i Volcanoes' Mauna Loa Trail, and *G. cuneatum ssp. tridens,* is common above the Haleakalā entrance and on the south crater wall. Both forms grow to be three- foot high shrubs with oblong leaves which have three to six conspicuous teeth at the tip. Both bear white flowers with purplish veins. In the Hawai'i plant, the leaves are green above and silvery beneath; in the Maui subspecies the leaves are silvery on both surfaces.

G. multiflorum is a woody shrub in Ko'olau Gap of Haleakalā where it is sustained by the frequent fog and rains which sweep up the Gap. Its leaves are ovate, pointed at the tip, with several teeth along each side. *G. arboreum* is a very rare tree with red flowers that grows on the outer slopes of Haleakalā.

Cranesbill, *G. homeanum,* is an alien, pink-flowered, low-growing herb on Haleakalā's crater floor and in Hawai'i Volcanoes' Kīpuka Puaulu. It is native to Australia and New Zealand.

Spurge Family

Euphorbiaceae
Aleurites moluccana;
kukui, candlenut

The kukui was a valuable tree used by the Hawaiians for illumination, dyes, medicine, tanning, ornamentation, and other purposes. It was probably brought to Hawai'i by early Polynesian settlers. Now, it is a common tree in moist lowland forests, up to about the 2,000 foot elevation. The trees in the coastal section of Hawai'i Volcanoes and the large groves on Hōlei Pali can be identified from a distance by their very light green foliage. It also grows at Haleakalā's 'Ohe'o Gulch.

The leaves are light-colored because they are covered with tiny whitish hairs. The leaves are long-stalked, with blades about eight inches long, and are often three-lobed and somewhat resemble maple leaves. Large clusters of small white flowers produce nuts which are about two inches in diameter. The outer part of the fruit, a green husk, later splits revealing a large black seed.

The kernel of the seed is rich in oil, and nuts were strung on midribs of coconut leaflets, then burned as candles. The oil was also pressed from the seeds and burned in stone lamps. Raw nuts are a powerful laxative, but roasted nuts are tasty. They are often pounded and mixed with salt and chili peppers to form 'inamona, a relish. Gum from the bark of the tree was used to strengthen bark cloth and to preserve fish nets. Occasionally canoes were made from the soft, light wood. Dyes were obtained from the bark of stems and roots, from the husks, and from the shells of the nuts. Several medicines were obtained from different parts of the plant. The black shells of the nuts were strung into leis, a custom that continues today.

This is the state tree of Hawai'i.

Bayberry Family

Myricaceae
Myrica faya; firetree

The firetree, native to the Canary Islands, Madeira, and the Azores, was introduced to Hawai'i in the 1920s, probably by Portuguese immigrants, who apparently made wine from the fruit and used the tree as a source of firewood. It grows to be a tree up to 45 feet tall, with smooth leathery leaves up to one inch wide and four inches long. Small, inconspicuous flowers are borne on finger-shaped spikes among the leaves. The fruits are small, dark red or black, slightly fleshy and the fruiting spikes are quite conspicuous. The fruits are eaten by birds, particularly the

introduced white-eye, and occasionally by pigs, and spread widely. This alien plant is now an important pest in several places in the islands, but particularly so in Hawai'i Volcanoes.

The firetree gets an early start in areas subject to volcanic activity, in part because it is able to fix nitrogen. It grows quickly into a large tree which casts very dense shade that inhibits the growth of native plant species. Currently a search is being made in its original homelands to locate organisms which may be useful for its biological control.

Mango Family

Anacardiaceae
Schinus terebinthifolius;
Christmas berry, wilelaiki

Christmas berry is a shrub or small tree introduced to Hawai'i from Brazil. It may be found in the coastal district at Hawai'i Volcanoes and at Haleakalā's 'Ohe'o Gulch.

Plants have compound leaves with five to nine leaflets, each leaflet one to three inches long, one-half inch wide. White flowers at branch tips develop into dense clusters of small bright red fruit, which are used as decorations, particularly at Christmas. Wilelaiki is the Hawaiian name for Willie Rice.

He was a politician who wore a lei of Christmas berry fruits as a hat band.

Christmas berry often grows in dense thickets, crowding out other plants, so it's one of the serious weed pests in Hawai'i.

The fruits today are marketed as "pink pepper," a gourmet spice. This spice should be used with caution as it may cause severe reactions in people who are sensitive to other plants in the *Anacardiaceae* such as mango, poison ivy, or poison oak.

Holly Family

quifoliaceae
x anomala; kāwa'u

e kāwa'u is a medium-sized
ee which is frequent in rain
rests such as the area around
urston Lava Tube. A fine
and grows at Palikū in
aleakalā Crater.

e oval, leathery leaves are a
ining dark green above and
mewhat lighter beneath. These
ually have smooth edges, but
aves on seedlings often have a
w teeth on the margins. The
aves never have the sharp
iny margins of English or
nerican holly, which may
flect the fact that the kāwa'u
olved in Hawai'i, where there
ere no browsing mammals until
ey were introduced by humans
er the past two centuries.
owers are small and white, in
usters among the leaves. Fruits
e berry-like, black, and about
ie- fourth inch in diameter.

e wood is white and quite
ft. It has been used to make
ddle trees.

Soapberry Family

Sapindaceae
Dodonaea, Dodonaea viscosa;
'a'ali'i, hopseed

46 'A'ali'i are common in both parks, except in dense wet forests. They occur in a bewildering variety of forms which may represent either varieties or distinct species. Near the coast in Hawai'i Volcanoes they are shrubs a few feet high with hairy leaves. At 5,000 to 6,000 feet on the Mauna Loa Road they become trees 20 to 25 feet tall, while at higher elevations they are again small shrubs. In Haleakalā they are shrub-size along the highway and in the western end of the crater, but tree-size at Palikū and Kaupō Gap.

The leaves are elongate, narrow, and often spatulate. Flowers develop in inconspicuous clusters at branch tips. The plants usually bear either female or male flowers. However, in many places male plants produce a few bisexual flowers toward the end of the flowering season, which then form a few fruits. Thus some plants bear many fruits, others only a few.

The fruits are conspicuous and attractive capsules, each about one-half inch in diameter, and with two to four broad wings. In most plants the capsules are shades of red, ranging from pink through scarlet to maroon, but some plants have yellowish green or brownish green fruits.

The Hawaiians prepared a red dye from the capsules by boiling them in a calabash of water to which hot stones were added. The wood is extremely hard and durable, and was used for such items as spears. When trunks of sufficient size were available they were used for house timbers. In modern times the capsules and leaves are used for lei-making.

Soapberry Family

Sapindaceae
Sapindus saponaria; aʻe, mānele

The aʻe is a large tree which is abundant in Kīpuka Puaulu and Kīpuka Kī. In Hawaiʻi it is known only from Hawaiʻi Volcanoes and adjacent areas, and from the Kona section of the island of Hawaiʻi. It is known from Mexico to South America, and from a number of Pacific islands as far west as New Caledonia. It also occurs in Africa. It is called aʻe in the park and mānele in Kona.

Trees may reach eighty feet in height and three feet in diameter. The leaves are six to ten inches long and are divided pinnately into six to twelve leaflets. Although the species is reported to be evergreen in tropical America, in Hawaiʻi the trees are deciduous. Each tree loses its leaves for about a month sometime between January and April. After new leaves develop, clusters of small flowers form at branch tips. Some of the flowers develop into round fruits about one inch in diameter, each with a large round black seed embedded in a fleshy pulp.

The seeds are occasionally used today in making seed jewelry, but there are no other recorded Hawaiian uses of the plant. The fruit pulp contains a soap-like substance, saponin, and is used in Mexico as a shampoo.

Mallow Family

Malvaceae
Sida fallax; 'ilima

There are many forms of 'ilima in Hawai'i, but the common ones in Hawai'i Volcanoes are small prostrate shrubs in exposed areas along the coast, and erect shrubs two to four feet tall from the lowlands up to about 6,000 feet. All have downy, grayish green heart-shaped leaves one to one- and-one-half inches long. Flowers resemble small hibiscus blossoms, usually yellow to orange, about one inch in diameter. They open early in the morning and wither by afternoon.

Hawaiians cultivated certain forms of 'ilima for their flowers which were strung to make leis for special occasions. Other forms were cultivated for their longer stems which were used as slats in building grass houses.

Mallow Family

Malvaceae
Thespesia populnea; milo

Milo is a tree reaching 20 to 40 feet in height. It may have reached the Hawaiian Islands by natural means of dispersal or it may have been introduced by early Polynesian immigrants. It grows today primarily in coastal regions, although it can grow some distance inland. It is common along the coastline in the Kalapana section of Hawai'i Volcanoes, and in Haleakalā at 'Ohe'o Gulch.

The leaves are heart-shaped, about five inches long, and bright glossy green. Flowers resemble small hibiscus; they are light yellow with deep purple centers when they open in the morning, but the yellow gradually changes to purplish pink as the day progresses. Seed pods are brown, five-sided, about one inch wide and one-half inch long.

The tree was planted because it was a useful shade tree in the hot, coastal areas, and because its dark wood was highly prized for calabashes.

Mallow Family

Malvaceae
Hibiscus tiliaceus; hau

Hau is by nature a lowland tree, most common in the Kalapana section of Hawai'i Volcanoes and the 'Ohe'o Gulch area of Haleakalā. It has also been planted near the Volcano House. It is probably indigenous to Hawai'i and to most other tropical areas, and was likely spread by floating seeds, but it may have been brought to Hawai'i by early Polynesian settlers.

Sometimes plants grow as upright but crooked trees to be about 20 feet tall. More often plants have long, spreading branches, which form dense, almost impenetrable tangles. Leaves are broadly heart-shaped, from two to ten inches across, deep green above and whitish below. Flowers are bright yellow when they open, sometimes with a reddish brown "eye" in the center. As the day progresses the flower becomes reddish or purplish brown, and drops off during the second day.

The fibrous bark was an important source of cordage, and was also used to make a coarse grade of kapa (tapa). The wood is light but strong and so it was used for the struts that attached the outriggers to the canoe hull, and occasionally for the outrigger itself. It was also used for adz handles. A block of soft hau wood, in combination with a pointed stick of a hard wood such as olomea (*Perrottetia sandwicensis*), was used as a fire-making tool.

'Ākia Family

Thymellaceae
Wikstroemia phillyreifolia,
W. sandwicensis; 'ākia

At least 12 species of 'ākia are known from the Hawaiian Islands. They are all similar, and difficult to distinguish, differing mainly in minor details of flower structure. Plants occur here and there at lower and middle elevations in both parks, and *W. phillyreifolia*, below, is quite common in the Kalapana section of Hawai'i Volcanoes and near Kīpuka Nēnē. *W. sandwicensis*, above, is a shrub or small tree in the wet forests around Thurston Lava Tube and in the 'Ōla'a tract.

The plants are shrubs, or occasionally small trees, with leathery leaves from one-half to four inches long. Small, fragrant yellow flowers are borne in clusters and are followed by juicy red or orange fruits from one-fourth to one-half inch in diameter.

The bark of 'ākia contains strong fibers which were used for cordage, and Hawaiians used the pounded plants as a narcotic fish poison. They also used the plants medicinally, as a laxative and as an asthma treatment. Some species of *Wikstroemia* in other parts of the world are extremely poisonous, and some writers indicate that 'ākia was used as one component of a poison drink for the execution of criminals in Hawai'i. However, its effects are not verified, and some people have eaten 'ākia fruits without ill effects. This practice is not recommended.

Myrtle Family

Myrtaceae
Psidium guajava; guava, kuawa

A native of tropical America, the guava was introduced to Hawai'i about 1800, and has become widely naturalized. In Hawai'i Volcanoes it grows from sea level to about 4,000 feet and is especially common in the Kalapana section and near Kīpuka Nēnē. It is common at Haleakalā's 'Ohe'o Gulch.

In wet forest areas the guava becomes a tree twenty to thirty feet tall, but in exposed areas it is a shrub six to ten feet high. The bark is a smooth, reddish brown, with a mottled greenish pattern. Leaves are in pairs, four to six inches long on short leaf stalks. White flowers are borne singly in leaf axils, and are one to one-and-one-half inches in diameter with many stamens. Fruits are about two inches in diameter, yellow on the outside, and pink or white internally with many brown seeds.

The fruits are edible and form the basis of juice and jelly industries, but the wild plants have little use as food, and they crowd out many desirable native species.

Myrtle Family

Myrtaceae
Psidium cattleianum;
strawberry guava, waiawī

The strawberry guava, a native of the American tropics, is one of the most serious alien plant pests in Hawai'i today. Plants become trees twenty to twenty-five feet tall, with paired, thick, leathery leaves, two to five inches long. Small white flowers are followed by many-seeded, fleshy fruits which are almost spherical in shape and about one inch in diameter. In one form the ripe fruits are red to purplish red when ripe, in another they are yellow. The fruits are tasty when fresh, and make excellent jams and jellies.

However, the fruits are also eaten by pigs and birds, which help to spread the seeds into native forests. Strawberry guava plants grow thickly, and cast such dense shade that few native plants can compete successfully with them. Rather quickly the understory plants in 'ōhi'a forests can be replaced by strawberry guava, and when eventually the

'ōhi'a trees forming the canopy die, the area is converted to a strawberry guava forest with very low biological diversity. Strawberry guava, and the pigs which help spread it, are particularly important pests in Kīpahulu Valley in Haleakalā National Park, and in wet forests below about 4,000 feet elevation in Hawai'i Volcanoes. In these places the National Park Service is making major efforts to control these aliens.

Myrtle Family

Myrtaceae
Metrosideros polymorpha;
'ōhi'a lehua

This is the most common and conspicuous tree in both parks. It grows from near sea level at Kalapana to tree line at about 8,200 feet on Mauna Loa. In Haleakalā it is abundant above Park Headquarters and at Palikū. It is the pioneer tree on new lava flows and is the dominant tree in the wettest rain forests. On dry lava flows it is frequently a shrub less than 10 feet tall; in somewhat wetter areas, as along the Chain of Craters Road, it is a straight tree 20 to 30 feet tall. In Kīpuka Kī and Kīpuka Puaulu, as well as in wet forests, it can reach a height of sixty to eighty feet and a diameter in excess of four feet. In extremely wet mountain bogs, as in the Alaka'i Swamp on Kaua'i, it is a dwarf shrub only a few inches high.

The leaves grow in pairs along the stem, usually on short stalks, and may be smooth and green on both surfaces, or may have a dense coating of whitish or rusty hairs on the underside. The young leaves are sometimes red, sometimes green, sometimes smooth, sometimes hairy. The flowers grow in clusters, with long protruding stamens, giving the cluster the appearance of a powder puff. These are usually scarlet, but can be salmon, orange, yellow, chartreuse, or white. The species name *polymorpha* is particularly appropriate. Native birds such as the 'apapane and 'i'iwi feed on the nectar produced by the flowers, and assist in pollination. The tiny windblown seeds are produced in small woody capsules.

The wood is hard, heavy, and dark red, and was used by Hawaiians to make spears, mallets, and images. Today it is used as flooring, for keel blocks of ships, and for a few other special uses where very hard wood is required.

Ginseng Family

Araliaceae
Cheirodendron trigynum; 'ōlapa

The 'ōlapa is a medium-sized tree which is frequent in wet forests. It is seen easily at Hawai'i Volcanoes' Thurston Lava Tube and Kīpuka Puaulu. It grows within Haleakalā Crater, around Palikū, in Kīpahulu, and at 'Ulupalakua, outside Haleakalā.

The leaves are in opposite pairs, with each leaf bearing three to five leaflets in a palmate or finger-like arrangement. Small green flowers form in clusters at branch tips and develop into black fruits about one-fourth inch in diameter.

In ancient hula performances the younger and more active dancers are placed in a group called 'ōlapa, which takes its name from the motion of the leaves swaying gracefully in the slightest breeze. A bluish dye can be obtained from the leaves and bark.

Heath Family

Ericaceae
Vaccinium reticulatum; 'ōhelo
V. calycinum; 'ōhelo kau la'au

The 'ōhelo is a shrub about two feet high which is especially common in Kīlauea Crater and near the Hawaiian Volcano Observatory. In Haleakalā it is common along the road above Park Headquarters, and is scattered across the crater. Its leaves are leathery, oblong to nearly-circular in shape, about one inch long, and often bluish or grayish green in color. The flowers are usually red, and are followed by red, orange, or yellow berries one-fourth to one-third inch in diameter. The 'ōhelo is a member of the same genus as blueberries and cranberries, and like them, its juicy, slightly sweet fruits are edible both raw and cooked.

The Hawaiians considered 'ōhelo sacred to Pele, the volcano goddess. When visiting Kīlauea they would pick a branch bearing berries and throw it into the molten lava as an offering to Pele before eating any of the berries themselves.

In the rain forest around Thurston Lava Tube is a closely related species, *V. calycinum,* 'ōhelo kau la'au or tree 'ōhelo. This plant is a shrub six to ten feet tall with bright green thin leaves and scarlet berries which have a somewhat bitter taste. At elevations above 6,000 feet on Mauna Loa is a variety of 'ōhelo, formerly called *V. peleanum* in which the flowers and fruits are usually dark purple. The 'ōhelo with purple fruits along Haleakalā's Halemau'u Trail and in Ko'olau Gap is yet another variety formerly called *V. berberifolium.*

Epacris Family

Epacridaceae
Styphelia tameiameiae;
pūkiawe

These distinctive shrubs occur throughout both parks, from near sea level to the upper limits of shrub growth. Within Hawai'i Volcanoes they are most conspicuous in the open shrubland along the Mauna Loa Road, between Kīlauea Military Camp and the Volcano Observatory, as well as on the upper slopes of Haleakalā along the road, and within Haleakalā Crater. They also grow in a variety of habitats, including rain forests. *S. tameiameiae* grows from near sea level to more than 6,000 feet on Mauna Loa and Haleakalā. This species also is native in the Marquesas Islands.

Pūkiawe plants are stiffly-branching shrubs with small, leathery, oblong, sharply- or bluntly-pointed leaves. These are one-fourth to one-half inch long and whitish on the lower surface. The white flowers, which develop in leaf axils, are small and inconspicuous. The berry-like fruits, up to one-fourth inch in diameter, vary in color from plant to plant and may be white, pink, scarlet, or maroon. Although they look appetizing, they are dry, mealy and lack any taste.

Hawaiians used the brightly-colored fruits in leis. The wood is said to have been used in cremating the bodies of outlaws. In ancient times the kapu system prevented chiefs from mingling freely with the common people, but the kapu could be removed temporarily if the chief exposed himself to the smoke from burning pūkiawe.

Myrsine Family

Myrsinaceae
Myrsine lessertiana;
kōlea lau nui

Kōlea lau nui is endemic to the Hawaiian islands, where it is a common tree 30 to 40 feet tall in wet forests. In Hawai'i Volcanoes it is most easily found in the forest near Thurston Lava Tube. In Haleakalā it is common around Palikū and in Kaupō Gap.

Young leaves are sometimes pale green, but more often are pink. Mature leaves are deep green and leathery. Small green flowers occur among the leaves and on older parts of the branches where leaves have already fallen. The ripe fruits are purple to black berries.

The hard wood was used by the Hawaiians for house beams and posts, and as anvils for beating kapa (tapa or bark cloth). A red dye was obtained from the bark, and a black dye was derived from the charcoal of the burned plant.

Ebony Family

Ebenaceae
Diospyros sandwicensis; lama

The lama is a dominant tree in dry forests at lower elevations. It is abundant in the Kalapana section of Hawai'i Volcanoes.

The trees have small leathery leaves, dark green when mature but usually reddish when young. Tiny unisexual flowers form in leaf axils, with each tree bearing only male or female flowers. After pollination the female flowers develop into juicy fruits about one inch long and one-third inch in diameter.

The fruits are first green, then yellow, then bright red when they are ripe. When fully ripe the fruits have a pleasantly sweet taste, but when unripe they are extremely astringent, like those of their close relative, the persimmon.

The wood is hard and light in color. This was a sacred wood used in the construction of temples, and also had a special use with the hula. During hula performances a block of lama wood, wrapped in yellow bark cloth, and placed on a special altar, was used to symbolize Laka, the goddess of the hula.

Morning Glory Family

Convolvulaceae
Ipomoea indica;
koaliʻawa, koaliʻawahia,
morning glory
I. pes-caprae; pōhuehue,
beach morning glory

This attractive koaliʻawa, right, grows throughout Hawaiʻi Volcanoes in open areas from sea level to about 5,000 feet. It can be found readily in Kīpuka Puaulu.

The plant is a vine with silvery green, heart-shaped leaves four or five inches long. The tubular flowers, up to three inches in diameter, are light bluish violet when they open in the morning. During the day they gradually change to a deep pink and close in the evening.

The plant is reputed to have great medicinal value. Roots, stems, and leaves were used as laxatives, and crushed stems and roots formed part of a poultice applied to bruises and even to broken bones. The vines, which are quite strong, were also used as swings.

A related species, *I. pes-caprae*, below, the pōhuehue or beach morning glory, is common on sandy beaches throughout the Hawaiian Islands. It forms thick carpets in several places on Kīlauea's southern coast. It has purplish red flowers and leathery leaves which are cleft at the tip, resembling the shape of a goat's foot.

61

Verbena Family

Verbenaceae
Lantana camara; lantana, lākana

Lantana, originally brought to Hawai'i as a garden ornamental in 1858, escaped from cultivation and has become a major pest at elevations up to 4,000 feet. It is common throughout the Kalapana section of Hawai'i Volcanoes and at Haleakalā's 'Ohe'o Gulch.

The plant is a shrub, usually three to six feet tall but under favorable growing conditions it is much larger and has very prickly stems and branches. Flowers form in flat-topped heads two to three inches in diameter and are yellow or orange when they first open, but they later become pink, red, lavender, or white. Since flowers of many ages occur together, each head is multi-colored. Its black fruits are dispersed by birds.

Since lantana has become a serious weed in agricultural and pasture lands many efforts have been made to eradicate it, or at least to control it. Many insects have been brought to Hawai'i as biological control agents because they feed on lantana, and most plants are infested by one or more of these insects. Yet lantana continues to thrive in the islands.

African Violet Family

Gesneriaceae
Cyrtandra platyphylla; 'ilihia

Over 50 species of *Cyrtandra* are native in Hawai'i. These show much variation and some scientists believe there may be more than 150 native species. All these are shrubs which grow in wet forests, with many confined to a single valley or mountain peak. *C. platyphylla*, right, is the common species around Thurston Lava Tube.

It is a shrub two to ten feet tall, with opposite, hairy, heart-shaped leaves about six inches long. The flowers are white, about one inch long, and the fruits are large waxy white berries.

Coffee Family

Rubiaceae
Coprosma ernodeoides;
kūkaenēnē, leponēnē
C. menziesii, C. montana,
C. ochracea, C. rhynchocarpa,
C. pubens; pilo

There are six species of *Coprosma* growing within Hawai'i Volcanoes and five within Haleakalā. All have small inconspicuous flowers, with male and female flowers growing on separate plants. The female plants are conspicuous when bearing brightly-colored fruits resembling coffee berries.

Kūkaenēnē, left, is a low trailing shrub from open lava flows at higher elevations. The leaves are narrow and linear, about one-half inch long and one-eighth inch wide. Its fruits are black, spherical, and about one-third inch in diameter. Kūkaenēnē is common along the upper part of Hawai'i Volcanoes' Mauna Loa

Road and across the floor of Haleakalā Crater. The fruits are a favorite food of the nēnē, the Hawaiian goose.

C. menziesii, below, and *C. montana* are shrub forms of

pilo which grow in open, moderately dry forests. Both have ovate or nearly circular leaves, which are leathery and up to one inch long, and orange or yellowish orange fruits. *C. menziesii* grows along Hawai'i Volcanoes' Chain of Craters Road near Pauahi Crater, whereas *C. montana* grows above 6,000 feet on the Mauna Loa Road. It is also common across Haleakalā Crater and above park headquarters outside the crater. In Haleakalā's Kaupō Gap, *C. menziesii* is common.

C. ochracea, above, and *C. rhynchocarpa* are tree forms of pilo which grow in closed forests of Hawai'i Volcanoes. Both have thin oval leaves up to four inches long and one inch wide. *C. ochracea* has red or reddish orange, egg-shaped fruits about one-fourth inch long. It is common in the rainforest around Thurston Lava Tube. *C. rhynchocarpa* has scarlet egg-shaped fruits one-half inch long, tipped with a fleshy scarlet beak one-half inch long. It is abundant in Kīpuka Puaulu and Kīpuka Kī.

Coffee Family

Rubiaceae
Hedyotis centranthoides;
Kīlauea hedyotis

This is a shrub with weak, sprawling, almost vine-like branches which is common both in wet forest and on fairly recent open lava flows in the Kīlauea area. It is uncommon in Haleakalā's Koʻolau Gap. Pairs of oval leaves, two to three inches long, are widely scattered on the stems. Tiny yellowish green flowers grow in several clusters arising from the bases of the uppermost leaves, and develop into seed pods about one-fourth inch long.

This species is endemic to the Hawaiian Islands, and grows on all the main islands except Niʻihau and Kahoʻolawe. Although it is fairly common and a conspicuous plant, no Hawaiian name has been recorded for it. Perhaps it was not used by the Hawaiians.

Coffee Family

Rubiaceae
Canthium odoratum; alahe'e

The alahe'e, a tree found in dryer lowland areas, is an indigenous species, native not only to Hawai'i but to many other Pacific Islands. It is fairly common in the dryer lowlands.

Trees are small, usually ten to fifteen feet tall, with shining dark green oblong leaves about three inches long. Clusters of fragrant white flowers develop at branch tips and these later form small black fruits.

The alahe'e is a useful ornamental plant. Its evergreen foliage is especially attractive, and the fragrant white flowers are an extra bonus for a month or two each year. The wood was used by the Hawaiians for digging sticks, and a black dye was obtained from the leaves.

Coffee Family

Rubiaceae
Morinda citrifolia; noni,
Indian mulberry

The noni, a small tree growing on lava rocks near the coast, was probably introduced to Hawai'i by early Polynesian settlers. It may be found in coastal areas of Hawai'i Volcanoes and is common near Haleakalā's 'Ohe'o Gulch. Its leaves are opposite, ovate, and often 8 or 10 inches long. White flowers develop in small round heads, and the head later grows into an irregular, light yellow, compound fruit about the size and shape of a potato.

The Hawaiians obtained a yellow dye from the bark of the stem and a red dye from the root bark. In times of famine the rather badly smelling fruits were eaten, but the fruits, leaves, and bark were used mainly to treat a variety of ailments.

Coffee Family

Rubiaceae
Psychotria hawaiiensis;
kōpiko ʻula

There are several kinds of kōpiko in Hawaiʻi, and this species is a very common understory tree in Hawaiʻi Volcanoes' Kīpuka Kī and Kīpuka Puaulu, and in ʻōhiʻa lehua forests throughout the island of Hawaiʻi.

Mature trees are 30 to 40 feet tall, with pairs of dark green, shiny leaves, broader nearer the tip than the base. Small, white flowers grow in open clusters at branch tips, and later these form fleshy orange berries one-half inch long with two large seeds.

Naupaka Family

Goodeniaceae
Scaevola taccada;
naupaka kahakai, huahekili
S. kilaueae; huahekili uka
S. chamissoniana;
naupaka kuahiwi

There are three species of naupaka that grow in Hawai'i Volcanoes, two in Haleakalā. All have flowers which split open down one side only, giving rise to the common but mistaken impression that the flowers are only "half-flowers."

Naupaka kahakai, right, is a shrub up to eight feet high growing near the coast. Look for it along the sea cliffs at Hawai'i Volcanoes and at 'Ohe'o Gulch in Haleakalā, where it grows as a sprawling shrub three to six feet tall. Each branch tip bears a thick cluster of shiny, somewhat succulent leaves. The flowers are white, sometimes with purplish streaks, in small clusters among the leaves. The white fleshy fruits, which gave rise to the name "huahekili" (hailstones), float on the sea, and have aided in spreading the species. It grows on nearly all tropical Pacific islands.

Huahekili uka, above, is an open sprawling shrub, usually one to two feet tall, which grows in open areas along the Chain of Craters and Hilina Pali Roads. It has yellowish white flowers and black fruits which are dispersed by birds. As the scientific name implies, this species is restricted to the Kīlauea region; it grows nowhere else in the world.

Although it is locally common in a few spots, its distribution is extremely limited, in an area subjected to volcanic activity.

Naupaka kuahiwi is an upright shrub up to eight feet tall, with white flowers and black fruits. It grows in wet forests such as those near Thurston Lava Tube, in the 'Ōla'a Tract, and in Haleakalā's Kaupō and Ko'olau Gaps. It is found in similar habitats on most of the main Hawaiian Islands.

No Hawaiian uses for these plants have been recorded, although naupaka kahakai is today grown widely as an ornamental plant.

Sunflower Family

Compositae
Dubautia ciliolata, D. scabra,
D. menziesii,
D. platyphylla; kūpaoa

Several species of kūpaoa occur in both parks. Common in dry forests is *D. ciliolata*, left, an erect branching shrub two to three feet tall. Its leaves are about one-half inch long, stiff, pointed upward along the stem, crowded, and usually borne in whorls of three. Small yellow nodding flower heads form at branch tips. Common along the Devastation Trail is *D. scabra*. This has branches which trail along the ground and are erect only at the tips, except for flowering branches which may be two feet tall. Its leaves are borne singly, widely separated from one another, are up to one inch long, and point outward or even toward the base of the stem. The small grayish white flower heads are erect.

At Haleakalā, *D. platyphylla* grows mostly within the crater.It has leaves two to four inches long and one-half to one-and-one-half inches wide with seven to eleven tall prominent veins. *D. menziesii* grows both within the crater and on the rim, up to the 10,000 foot elevation. It has leaves one to two inches long and one-fourth to one-half inch wide with three (or rarely five) prominent veins. Both plants are shrubs, reaching about nine feet in height.

The root of *D. ciliolata* is fragrant and was used by the Hawaiians to perfume bark cloth and featherwork.

Sunflower Family

Compositae
Argyroxiphium sandwicense;
'āhinahina, hinahina, silversword

The silversword, perhaps the best known of Hawai'i's native plants, is as famous as Haleakalā's crater. Its evolution has resulted in a spectacular plant unlike any others in Hawai'i. A dense rosette of leaves about two feet across protects the growing part of each plant, and silvery hairs on the leaves reflect the excess ultraviolet radiation. The reduced leaf size minimizes water loss, as do the hairs, which slow the desiccating wind.

Each plant grows as a spherical rosette of leaves for several years until it reaches maturity. Then, during the May to October flowering season, a single stem grows one to nine feet high and bears one hundred to six hundred yellow to maroon, nodding flower heads. After the seeds ripen the plant dies.

Silverswords were common across the floor of Haleakalā Crater until the early 1900s, when grazing goats and vandalizing travelers began to take their toll and threaten the species' existence. Under better protection now, silverswords are common in the crater, especially in dry, porous soil or volcanic cinders, as at Pu'u Nole. This is a high elevation plant, found between 6,000 and 12,000 feet. Other kinds of silverswords grow on Mauna Kea and Mauna Loa on the island of Hawai'i, and on West Maui's Mauna 'Eke and Pu'u Kukui. The Mauna Kea

silversword is considered to be a subspecies of the Haleakalā silversword, while the Mauna Loa plants are considered to represent a different species, and the West Maui plants still another species.

Photo Credits

In addition to the author, the following photographers contributed to this work:

Ed Bonsey
Brian Goring
Dwight Hamilton
Keith Hoofnagle
Glen Kaye
Margaret Linderer
George Linney
William Mull
Truman Osborn
Richard Rasp
James Sneddon
Gilbert Tanaka
Dan Taylor

Selected Readings

...or those interested in learning ...ore about Hawaiian plants, the ...llowing books are suggested:

...ARLQUIST, S. 1980. *Hawai'i, ...Natural History.* Second ...dition. Lawai, Hawai'i: Pacific ...opical Botanical Garden.

...JDDIHY, L. W. and C. P. ...TONE. 1990. *Alteration of ...ative Hawaiian Vegetation.* ...onolulu: University of ...awai'i Press.

...EGENER, O. 1945. *Plants of ...awaii National Park.* Reprint ed. ...n Arbor, Mich: Edward Bros. ...rst Published in 1930).

NEAL. M. C. 1965. *In Gardens of Hawai'i.* Rev. ed. B. P. Bishop Museum Special Publication 50.

SOHMER, S. H. and R. GUSTAFSON. 1987. *Plants and Flowers of Hawai'i.* Honolulu: University of Hawai'i Press.

ST. JOHN, H. 1973. *List and Summary of the Flowering Plants in the Hawaiian Islands.* Lawai, Kaua'i: Pacific Tropical Botanical Garden, Memoir No. 1.

STONE, C. P., and L. W. PRATT. 1994. *Hawai'i's Plants and Animals: Biological Sketches of Hawaii Volcanoes National Park.* Hawaii Natural History Association.

WAGNER, W. L., D. R. HERBST, and S. H. SOHMER. 1990. *Manual of the Flowering Plants of Hawai'i.* 2 vols. Honolulu: University of Hawai'i Press and Bishop Museum Press.

Index

76

Index

Notes

9 780940 2950

ISBN 0-94029 5-09-